Heitor Hermeson de Carvalho Rodrigues
Fábio Lavor Bezerra
Demócrito Sobreira da Cruz Cortez

Electromagnetism

Heitor Hermeson de Carvalho Rodrigues
Fábio Lavor Bezerra
Demócrito Sobreira da Cruz Cortez

Electromagnetism

from theory to experimental results

Imprint

Any brand names and product names mentioned in this book are subject to trademark, brand or patent protection and are trademarks or registered trademarks of their respective holders. The use of brand names, product names, common names, trade names, product descriptions etc. even without a particular marking in this work is in no way to be construed to mean that such names may be regarded as unrestricted in respect of trademark and brand protection legislation and could thus be used by anyone.

Cover image: www.ingimage.com

This book is a translation from the original published under ISBN 978-3-330-73818-8.

Publisher:
Sciencia Scripts
is a trademark of
Dodo Books Indian Ocean Ltd. and OmniScriptum S.R.L publishing group

120 High Road, East Finchley, London, N2 9ED, United Kingdom
Str. Armeneasca 28/1, office 1, Chisinau MD-2012, Republic of Moldova, Europe
Printed at: see last page
ISBN: 978-620-5-79329-9

Dedication

To my parents: Hernani Martins Rodrigues and Maria Helena de Carvalho Rodrigues and a friend-sister, Cleide Fernandes, *in memoriam*.

Author: Heitor Hermeson

"If A is success, then A equals X plus Y plus Z. Work is X; Y is leisure; and Z is keeping your mouth shut."

(Albert Einstein)

"Unshakeable faith is only that which can face reason head on, in all ages of humanity."

(Allan Kardec)

THANKS

Author: Heitor Hermeson de Carvalho Rodrigues

To God SU for his infinite love.

To my dear parents, for their great love and for always being present at the most important moments of my life.

To my dear friend-sister, Cleide Fernandes, *in memorium,* for the laughs, perseverance and always supporting my coming to Camocim and saying that I was going to make the difference in Science and Technology.

Finally, to all teachers, friends and colleagues at the University of Fortaleza (UNIFOR), specifically in the Experimental Physics Laboratory, and the Federal Institute of Education, Science and Technology of Ceará (IFCE) - Advanced Campus of Mombasa, who contributed directly or indirectly to the realization of this work, whose names it would not be possible to mention, so as not to commit injustice by omission, but whose memories I will cherish.

Author: Fábio Lavor

I thank in the first place God, my family, my true friends, my enemies, who taught me a lot, and the Spiritist Doctrine for having offered subsidies so that I could question myself and find within myself the reason for being reincarnated.

I thank the educational institutions through which I passed, such as UNIFOR, CENTEC and IFCE.

Autor: Demócrito Sobreira da Cruz Cortez

I thank first of all God, my children Ícaro and Ívina for the inspiration of every day, my wife Ingrid for the walk and complicity that led me to believe that everything was possible, my father and my mother, my family, my friends in the UFC.

I thank the educational institutions through which I passed, such as UFC and UECE.

I thank the IFCE for their support, especially my colleagues at the Advanced Campus of Mombaça who contributed directly or indirectly to the realization of this work, whose names I could not mention, so as not to commit injustice by omitting any, but whose memories I will cherish.

SUMMARY

1. INTRODUCTION

In the year of 2005, the authors (Heitor and Fabio) did the theoretical and practical discipline of Electromagnetism in the graduation course of Electronic Engineering in the University of Fortaleza (UNIFOR).

After the experience of this discipline aligned with the teaching course came the idea of gathering these works in a publication. After 2 years (2020) and with much effort and dedication, we present this book whose texts bring discussions based on the authors Halliday *et al* and Edminister who are experts in this area of knowledge.

The purpose of this book is to provide educational support to students of Technical Courses in Electronics, Electromechanics, Industrial Automation and Electrotechnics and Higher Education such as: Technologists and Electronic, Electrical, Mechatronic and Control and Automation Engineering, as well as technical reading for professionals and researchers.

The book is a complementary work to the classic ones so that the reader understands the application of theory in practice through tables, figures and graphs, besides containing summarized material for both electromagnetism professor and students and researchers. The texts are didactic, so that the reader will analyze and verify the main particularities of laboratory tests with the respective experiments and their justifications.

Keywords: Electromagnetism, Inductance, hysteresis.

CHAPTER 1: INDUCTANCE

1. 1Summary

This chapter describes the procedures and research required to prove the Faraday and Lenz laws, as well as the ideas of magnetic field lines, the explanation of the electromagnetic phenomena of self-inductance and mutual inductance, and the knowledge of the susceptibility and permeability properties of the medium in which these lines are inserted.

Inductors are electrical devices whose characteristic called inductance refers to the opposition to the variation of electric current or any variation in flux. There are several different ways to vary the magnetic or electric flux, in both cases an electric voltage is induced in the circuit and its value is equal in modulus to the variation rate of this flux. Thus, the essential elements are used to carry out the practice such as: inductor, billets, wires and equipment; oscilloscope, DC/AC voltage sources and milliamps.

In the experiments, the deflection in the milliampere is observed when introducing the magnet into the solenoid (synonym of inductor), as well as in the relative motion of both, as well as the increment of the electric current when introducing the billet inside the solenoid. It is concluded that the permanent ferromagnetic material exerts an electromotive force on conductive materials and the occurrence of the alignment of the magnetic domains in the iron billet when it is subjected to the magnetic field of the solenoid.

1. 2Introduction

It is known that the production of electric current requires the consumption of some form of energy, until mid-1831, a period in which magnetic and electric properties were considered separate phenomena and without any interference. However, the first study was carried out by physicist and chemist Hans Christian Oersted. Then, Michael Faraday discovered and explained the phenomenon of electromagnetic induction and later the scientist Heinrich Friedrich Lenz contributed to the understanding of the phenomenon [1-3].

Inductance is a characteristic of the inductor whose unit is Henry (H) and symbol (⟿) refers to the opposition to the variation of magnetic flux observed in part I of the experiment.

Magnetite is an element with its own magnetic properties, therefore independent of the medium in which it is inserted to produce a magnetic field. Analogous to the magnet used in Part I of the experiment.

Under these materials cited, one can vary the magnetic flux in different ways. In all these cases a voltage is induced in the circuit and its value is equal in modulus to the rate of change of magnetic flux $\frac{d\varphi_B}{dt}$ Faraday-Lenz's Law, $\oint E \cdot \mathrm{ds} = -\frac{d\varphi_B}{\mathrm{dl}}$ where the negative sign comes from Lenz's Law, which indicates the direction of the induced electric current as opposed to the variation of the magnetic flux which generated it, similar to the reference to the law of conservation of masses. Where, φ_B is the number of lines of the field that crosses the surface of the circuit. [3].

Part I inserts the billet, a ferromagnetic material, and has a strong exchange interaction, presenting a high magnetic susceptibility, where the application of a small magnetic field can result in a high degree of alignment of the magnetic moments [4].

As the direction of alignment varies from domain to domain the total magnetization of the billet may be zero, however when an external field is applied, the boundaries of the domains are directed such that the induced field in the material can be much larger than the initially applied field itself, hence, $H \gg M$ where H is the magnetic field and M is the magnetization, then $B = \mu_0 H + \mu_0 M (Wb.m^{-2})$, where $M = X_M H$, where $X_M = \mu_r - 1$, is the magnetic susceptibility.[3-4]

In Part II the phenomena of self-inductance and mutual inductance are studied. It is found that self-inductance depends on the geometric shape of the coil and is proportional to the rate of change of the electric current. As the inductor is integrated for the magnetic field, analogically, the capacitor is related for the electric field, in turn, the inductance only depends on geometrical factors $L = N \frac{\varphi}{i} e \ L = \frac{\mu N^2 A}{l}$[3].

In mutual inductance, the magnetic flux that passes through a circuit depends on the electric current in the circuit and on the electric currents in other circuits nearby, therefore it depends on the geometric arrangement of the two circuits, where $\varphi_M = L_2 I_2 + M_{21} L_1 (Wb)$[2-3].

1.3 Experimental Procedure

1.3.1 Part I

The following is a logical sequence in the assembly of the experiment with the respective materials made available in the laboratory:

A. The centre zero galvanometer is connected to the terminals of the 17.7 mH coil;

B. Hold the rod-shaped magnet by one of its ends and slowly approach, stop, and move it away from the centre of the inductor, observing what occurs in the galvanometer. Repeat this procedure by reversing the polarity of the magnet and changing the speed at which the magnet approaches/falls away from the coil. As shown in the diagram in Figure 1.1.

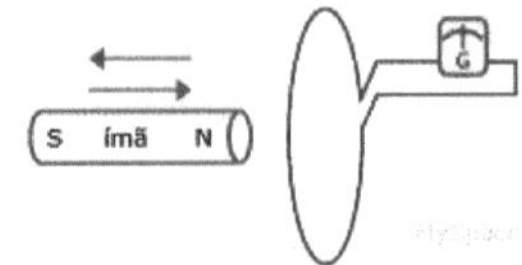

Figure 1.1 - Diagram of the magnetic field lines. Source: Own authorship, 2005.

C. One of the coils is connected to the 12/30V indication socket on the bench and the other coil is connected to the AC milliampere meter;

D. Move the inductors (without a plug in the core) closer together, and observe what happens. Move the inductors apart, insert the billets into the cores of the inductors, and bring them together again. As shown in Figure 2.1.

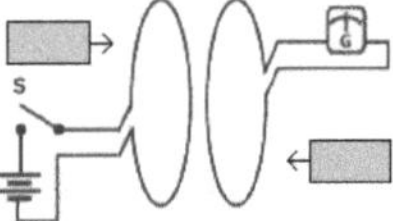

Figure 2.1 - Scheme of the variation of field lines with or without the introduction of billets. Source: Own authorship, 2005.

E. Using the mounting table with the protractor, connect the fixed coil to the function generator and set it to provide a sinusoidal signal, with a voltage of 12 Vpp and a frequency of 500 Hz. The moving coil will be connected in parallel to a 330 nF capacitor and to CH1 of the oscilloscope.

F. Determine the value of the induced voltage as a function of the angle, note the values and graphically represent these values.

1.3.2 Part II

G. Connect the generator to the coil terminals (600). Set a sinusoidal signal with 4 Vpp and a frequency of 1000 Hz.

H. The generator signal is measured.

I. Switch the oscilloscope probes to the other coil terminals (300).

J. The coil output signal is measured.

K. The billet is inserted into the coil.

L. Switch the oscilloscope probes to the coil terminals (600).

M. The generator signal is measured.

N. Switch the oscilloscope probe tips to the other coil terminals (300).

O. The coil output signal is measured.

P. Switch the oscilloscope probes to the coil terminals (600).

Q. The billet is removed from the coil.

R. The coil terminals (300) are short-circuited.

S. The generator signal is measured.

T. The billet is inserted into the coil.

U. The generator signal is measured. As shown in Figure 3.1:

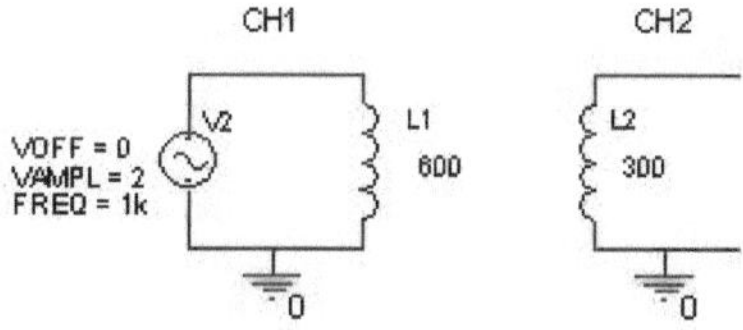

Figure 3.1 - Schematic of connections in the oscilloscope. Source: Own authorship, 2005.

1.4 Results and Discussions

1.4.1 Part I

By connecting the galvanometer to the terminals of the coil it is observed that the approach of the magnet there is a sensitization in the galvanometer. Ceasing the movement of the magnet, the passage of electric current ceases and when it inverts the poles of the magnet, perchance, reverses the direction of the electric current, where the magnet is called the inductor, the spiral is the induced and the current that is obtained is called induced current.

After connecting one of the coils (both without the billet) to the power supply and the other to the milliampere meter, and bringing the coils closer together, it can be seen that there is almost no sensitization in the milliampere meter, i.e., the induced current is minimal, but placing the billets and following the same previous procedure, it can be seen that there is a significant sensitization in the milliampere meter.

By Faraday's Law it is known that by obtaining a variation of magnetic flux inside a coil, it induces an electromotive force (e.f.m.) in this circuit [1-3.5]. Thus, propitiating an electric current in an inductor of "N" turns, a magnetic induction field (B) will be created, which being approached to the inductor that is connected in the milliampere meter, this will cause an e.e.m. and consequently an electric current will be observed. When the coils are without the billets and thus the inside of the coils has air, there is no significant magnetic permeability, therefore sensitization is not noticed in the milliampere meter (inexpressive mutual magnetic inductance).

By placing the billets, there is a magnetic susceptibility, facilitating a greater intensity of magnetic lines and mutual inductance are easily perceived. The electric current flow that varies with time creates an induced magnetic field that will affect the second coil and induce a current in it, so the billets help in a lower energy loss [1-3,5].

By varying the orientation of the position of the coil in space, the number of magnetic field lines crossing the surface of the coil varies, resulting in an induced voltage, as shown in Table 1.1, and transformed into a graph as shown in Figure 4.1, where, from a mathematical point of view, the curve obtained from the graph is a cosine.

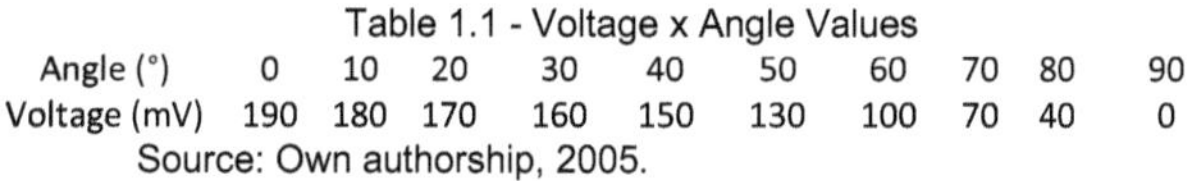

Table 1.1 - Voltage x Angle Values

Angle (°)	0	10	20	30	40	50	60	70	80	90
Voltage (mV)	190	180	170	160	150	130	100	70	40	0

Source: Own authorship, 2005.

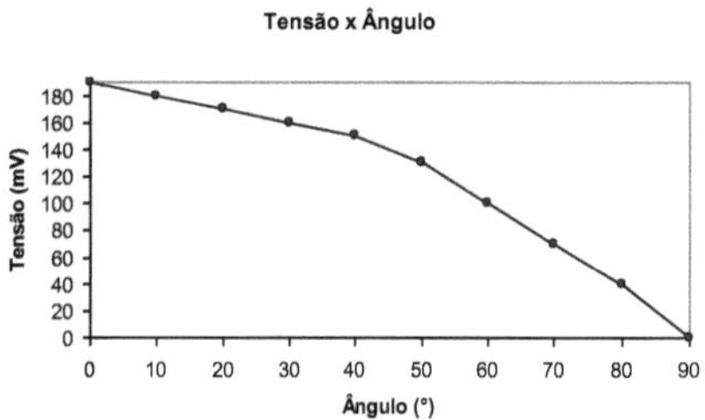

Figure 4.1 - Voltage x Angle graph. Source: Own authorship, 2005.

1.4.2 Part II

The values of the electrical voltage of the generator and the coil with or without billets are presented in Table 2.1.

Table 2.1 - Electrical Voltage Values

Without Targe (Volts)		With Tarugo (Volts)	
VG	$1,4 \pm 0,01$	VG	$3,2 \pm 0,01$
VB	$0,9 \pm 0,01$	VB	$1,4 \pm 0,01$
Short Circuit			
VG	$0,6 \pm 0,01$	VG	$0,6 \pm 0,01$

Source: Own authorship, 2005.

It can be seen that with billets the voltage increases, because there is a greater intensity of induced m.f.e. Although the phenomenon of mutual inductance occurs in both cases (without or with billet), with the billet there is less dispersion of the magnetic lines, and also the orientation of the ferromagnetic domains generates a higher flux density of the magnetic field lines.

When the 600 coil is short-circuited to the 300 coil, it is observed that there is no variation of the electric voltage when introducing or not the billets, although the phenomenon of mutual inductance occurs, the magnetic flux is continuous over a closed surface surrounding the 300 coil. Therefore, the magnetic flux density is equal to zero, that is, $\oint B \cdot ds = 0$ James Clerk Maxwell's magnetic field equation, as shown in Figure 5.1.

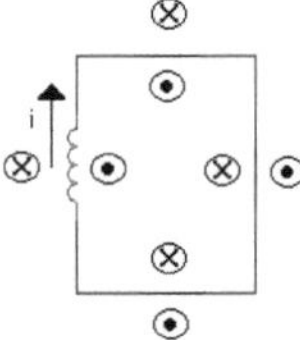

Figure 5.1 - B-field lines on 300 coil. Source: Own authorship, 2005.

1.5 Final Considerations

The existence of magnetic field lines is a characteristic of a permanent ferromagnetic material to exert an electromotive force on conducting materials when in motion.

The substantial increase in the total magnetic field, when the magnetic domains in the iron billet are subjected under the magnetic field of the solenoid fed by an alternating current, aligning with it.

The existence of an ordered magnetic flux when a force is exerted on the electrons in the conducting materials which makes them move in the opposite direction to the direction of the field which originated them. The self-induced voltage opposes the variation of the electric current, that is, it is a counterelectromotive force.

It has been proven that it is the variation in the magnetic field density rate that causes the phenomenon of magnetic induction to occur.

The magnetic flux, when two circuits are close together, depends not only on the electric current of one of the circuits, but also on the electric current of the nearby circuit, that is, the mutual inductance is the contribution of each circuit to the total field.

1.6 References

[1] TIPLER, P. A., **Física, eletricidade, magnetismo e ótica**, 4ª ed.

[2] RESNICK, R., HALLIDAY, D., WALKER, J., **Fundamentals of physics, electromagnetism,** 4th ed. LTC, Rio de Janeiro, 2000.

[3] HAYTT Jr, W. H., **Electromagnetism** 3a, pg.256, pg.258, pg.265-266, LTC, Rio de Janeiro, 1983.

[4] CALLISTER Jr, W., **Materials Science and Engineering an Introduction,** 4th ed. pg.659-662, pg.666, John Wiley, Canada, 1997.

[5] KRAUS, J. D., CARVER, K. R. **Electromagnetismo,** 2ª ed.

CHAPTER 2: AMPERE LAW AND TRANSFORMERS

2.1 Summary

The production of magnetic fields in conductors due to arbitrary currents can be analysed using Ampère's Law. Experiments were carried out with this knowledge and, having observed the induction effects, the principle of operation of the transformer was explained. Using the transformer with two wire windings surrounding the arms of a metallic frame and a current applied to one of the windings resulting in an induced voltage in the other winding while the magnetic flux is varying.

And when the metallic frame is opened, a decrease in the induced voltage is observed for the applied current. Therefore, the relations of voltage and electric currents between the windings vary according to the magnetic flux, and this, in the ratio between the number of winding turns, the different impedances, mutual inductance and saturation of the magnetic medium, demonstrating laboratory discrepancies in the test. Due to the closed metallic frame, the magnetic flux concentrates and reaches the secondary winding with few losses, then electromagnetic induction occurs and an electric current arises. Thus, Ampère's Law is verified in the experiment in the formalization of the relationship between electric current and magnetic field.

2.2 Introduction

In 1819, the physicist and chemist Oersted makes the discovery that an electric current passing through a straight wire produces a magnetic field with field lines formed by circles in perpendicular planes. The right hand rule is intended to be used as a north of the direction of the electric current, through the thumb, and the other fingers give direction of the magnetic induction field (B) [1]. A year later from Oersted's work, 1820,

Ampère standardizes the relationship between electric current and magnetic field, named for his name, called Ampère's Law $\oint \vec{H}.d\vec{S}=i_{envolvida}$ [2-4].

This law states that the line integral of the magnetic field H in any closed path is equal to the current involved by this path. [2] Around 1831, Michael Faraday and Joseph Henry independently and virtually simultaneously discovered electromagnetic induction, finding that an electric current could be generated magnetically, but that such an effect is observed only when the magnetic flux through the circuit varies with time, thus Faraday's Law, $f.e.m=-\frac{d\varphi}{dt}(Volts)$ [2-4].

Both observed that when an electric current varies in time, a magnetic field flows in the circuit which acts to induce an m.e.f. in this same circuit, the effects of which are opposite to the external m.e.f. which causes the current to vary in the first place [2-4].

They complemented their studies with the e.m.s. and currents induced in a coil (another synonym for inductor) by time-varying currents flowing in another coil nearby, and found that very strong induced e.m.s. could be excited in a coil having a significant number of turns of wire, by a smaller time-varying e.m.s. in a coil having relatively few turns.

In this way they built the first induction coils and invented the principles upon which the transformer operates. The principle of operation of a transformer is the variation of the modulus of B with respect to time, where an oscillating current is applied to a primary coil and a secondary coil is placed close to another, in general magnetically coupled by an iron core. As presented in Figure 1.2: [1]

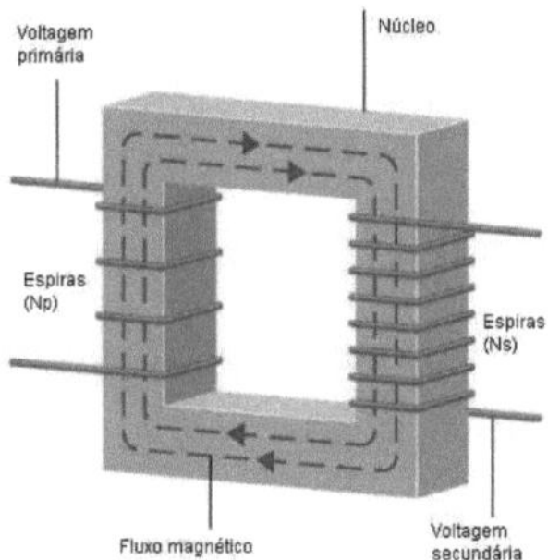

Figure 1.2. Diagram of a transformer where Np and Ns are the primary and secondary coils, respectively, and depend only on the geometry of the windings. Source: Own authorship, 2005.

2.3 Experimental procedure

2.3.1 Part I

The circuit in Figure 2.2 is assembled.

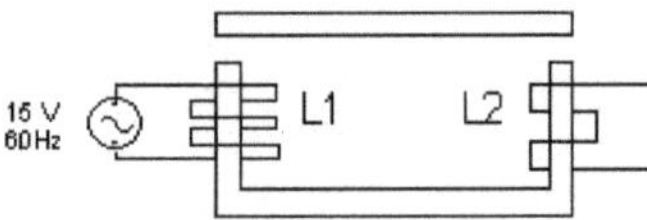

Figure 2.2 - Diagram of transformer assembly without the closing air gap. Source: Own authorship, 2005.

With the primary coil of 600 turns (L1), powered by a 15 V source$_{AC}$, and the secondary coil of 300 turns (L2), the voltage in coil L2 was measured. The air gap was then closed, as shown in Figure 3.2. The voltage in coil L2 was also measured.

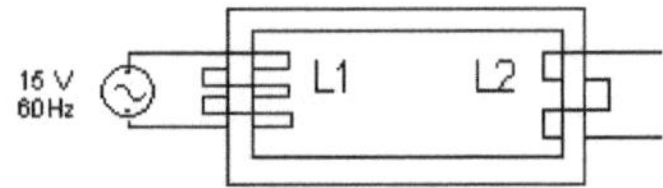

Figure 3.2 - Illustration of the transformer assembly with the closing air gap. Source: Own authorship, 2005.

The experiment was also performed using the 300 coil as primary (L1), powered by a 15 V source$_{AC}$, the voltage measurement is performed on the 600 coil (L2).

2.3.2 Part II

With the multimeter, the rheostats were adjusted so that the electric current in the circuit is 1 A, then a circuit was assembled with the two rheostats connected in parallel, as presented in Figure 4.2.

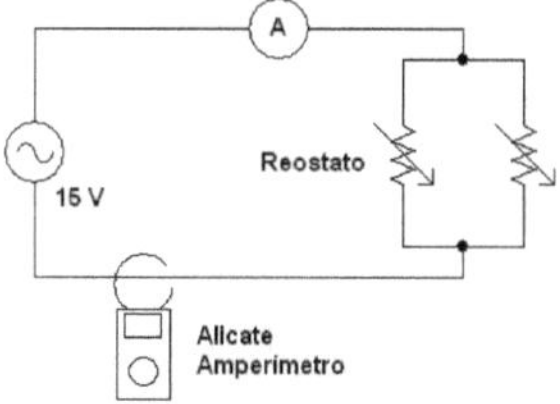

Figure 4.2 - Assembly scheme to perform the measurements. Source: Own authorship, 2005.

Electrical current measurements were performed in the circuit using the multimeter and the ammeter pliers, as shown in Figure 4.2. These measurements were also performed with one of the rheostats disconnected from the circuit and changing the AC source to DC.

2.3.3 Part III

We assemble the circuit of Figure 5.2 and carry out on the circuit electrical current measurements using the multimeter and the ammeter pliers. And electrical voltage measurements using the sensing coil, when it made an angle of 0° and 90° with one of the wires of the circuit and then with one of the rheostats turned off.

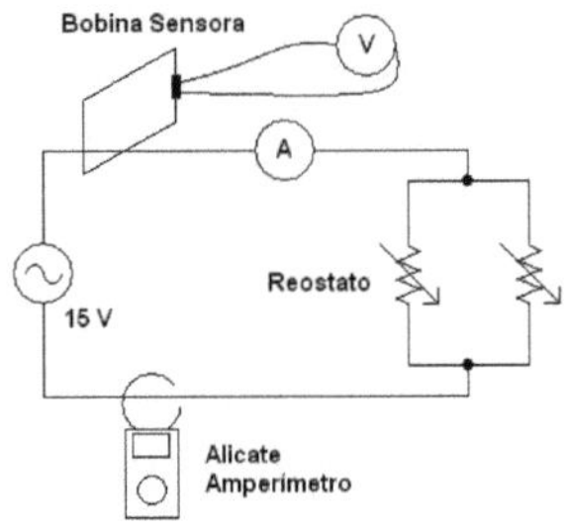

Figure 5.2 - Illustration of the experimental assembly. Source: Own authorship, 2005.

2.4 Results and discussion

2.4.1 Part I

The electrical voltage across coil L1 (600 turns) and L2 (300 turns) with the air gap open is shown in Table 1.2.

Table 1.2 - Measurements of the electrical voltages on the coils with the air gap open.

$V_{L1}(V)$	$V_{L2}(V)$
$14,5 \pm 0,1$	$1,3 \pm 0,1$

Source: Own authorship, 2005.

The electrical voltage across the same coils, with the air gap closed, is shown in Table 2.2.

Table 2.2 - Measurements of electrical voltages on the coils with the closing air gap.

$V_{L1}(V)$	$V_{L2}(V)$
$14,5 \pm 0,1$	$6,9 \pm 0,1$

Source: Own authorship, 2005.

The voltage on coil L1 (300 turns), powered by a 15 V source$_{AC}$, and L2, with open air gap, is shown in Table 3.2.

Table 3.2 - Measurements of the electrical voltages on the coils without the closing air gap.

$V_{L1}(V)$	$V_{L2}(V)$
$13,5 \pm 0,1$	$4,5 \pm 0,1$

Source: Own authorship, 2005.

The voltage on coil L1 (300 turns), powered by a 15 V source$_{AC}$, and L2, with the air gap closed, is shown in Table 4.2.

Table 4.2 - Voltage measurements on the coils with the closing air gap.

$V_{L1}(V)$	$V_{L2}(V)$
13,5± 0,1	28,1± 0,1

Source: Own authorship, 2005.

According to the electrical voltage values shown in tables 1.2 and 2.2, there is a voltage decrease relationship between V_{L1} e V_{L2}, that is, there is a voltage reduction behaviour, where V_{L2} presents a voltage lower than V_{L1}. However, between tables 1.2 and 2.2, there is also a difference in this relationship when the air gap is open or closed.

It happens that the magnetic flux in the closed air gap, there is no displaced liquid charge, so the charge carriers are not polarized, in other words, the magnetic circuit is closed, favoring with lower losses, justifying such difference [4].

According to the voltage values shown in Table 4.2, there is a voltage increase relationship between V_{L1} e V_{L2} that is, there is an increasing behavior in the electric voltage, where V_{L2} presents a voltage greater than V_{L1}. Knowing that in the coil V_{L1} we have : $V_{L1}=N_1\frac{d\varphi_1}{dt}$, and in the coil V_{L2}: $V_{L2}=N_2\frac{d\varphi_2}{dt}$.

And being $\varphi_1=\varphi_2$ implies that: $\frac{d\varphi_1}{dt}=\frac{d\varphi_2}{dt}$ .Thus, a general relationship is observed between V_{L1} e V_{L2} where an electric voltage rise or fall behaviour, and governed by the relation $V_{L2} = \left(\frac{N_2}{N_1}\right) \cdot V_{L1}$ The number of lines crossing the surface of the iron core, where a voltage is induced in the secondary coil and its value is equal in magnitude to this variation.

The electrical voltage values in Table 3.2 do not show the explicit behaviour of a voltage lift because of the displaced liquid charge, thus the charge carriers are polarised, i.e. the magnetic circuit is open, presenting higher losses [5].

If the electric current is varying, the flux through the coil is also varying, it is observed that there is an induced voltage only while the magnetic flux is varying (by means of the AC source), that is, when the flux is constant, as in the case of a DC source, the induced voltage in the secondary is zero, but soon after switching on, the current varies from zero until it reaches its finite DC value, so in this "variation" an instantaneous induced voltage occurs.

2.4.2 Part II

The current in the circuit of Figure 4.2, with the multimeter was $1.90A$ and with the pliers ammeter $1.87A$. The relative deviation was 1.6%. When one of the rheostats in the circuit was turned off, the value of the electric current with the multimeter was $1.02A$ and with the pliers ammeter, $0.97A$. The relative deviation was 4.9%. However, when you used a 15 V source$_{DC}$, the value measured with the multimeter was $2.25A$. and with the pliers ammeter, $0.00A$.

The relative deviation was 100%. The differences between the measurements of the multimeter and the clamp meter are due to the fact that the clamp meter is sensitized by the magnetic field of the wire. This magnetic field is embedded in a medium of permeability $\mu_0 = 1$(air) and the pliers claws form an arc of radius r when wrapping the wire, this involved current is calculated by Ampère's Law [2-4]:

$$\oint \vec{H}.\vec{d}\mathrm{L}=i_{envolvida}\,(A/m) \tag{2.1}$$

$$\vec{B}=\mu_0.\vec{H} \tag{2.2}$$

$$\int_0^{2\pi} \frac{B_\phi}{\mu_\phi}.rd\phi=i_{envolvida} \tag{2.3}$$

$$B_\phi = \frac{\mu_\phi.i_{envolvida}}{2.\pi.r} \tag{2.4}$$

On the multimeter there are no such factors, μ_0 e r. When one of the rheostats is removed from the circuit in Figure 4.2, the electric current decreases because the electric voltage is constant, the factors μ_0 e r become significant causing a relative deviation of 4.9% in relation to the multimeter measurement.

When using a 15 V source$_{DC}$, it is impossible to measure an electric current with the clamp meter, because although there is a flux φ there is no rate of change of this flux that passes through the arc of the pliers ammeter. By Faraday's Law we have [2-4]:

$$f.e.m_{alicate} = \frac{-d\phi}{dt} \tag{2.5}$$

$$\phi = constante; f.e.m_{alicate} = 0 \tag{2.6}$$

So the measured value is 0 because the magnetic flux is constant, i.e. it does not produce an electromotive force on the clamp meter.

2.4.3 Part III

The value of the electric current with the pliers ammeter was 1.98A. The electrical voltage of the sensing coil was 0.00V for an angle of 90° and 1.9mV for an angle of 0° between the coil and the wire. When removing one of the rheostats, the values were 0.00V for an angle of 90° and 1.1mV for an angle of 0° between the coil and the wire. The current that flows through the wire generates a magnetic field around it. When we approach the sensing coil making an angle of 0° with the wire, we have by Ampère's Law [2-4]:

$$\oint \vec{H}.d\vec{L} = i_{envolvida} \tag{2.7}$$

$$\int H_\phi.\vec{a}_\phi.rd\phi.a_\phi = i_{envolvida} \tag{2.8}$$

$$\vec{a}_\phi.\vec{a}_\phi = 1 \tag{2.9}$$

$$H_\phi = \frac{i_{envolvida}}{2.\pi.r} \tag{2.10}$$

When we approach the sensing coil making an angle of 90° with the wire, we have:

$$\oint \vec{H}.\vec{d}\mathrm{L}=i_{\text{envolvida}} \tag{2.11}$$

$$\int H_\phi.\vec{a}_\phi.\mathrm{r}\mathrm{d}\mathrm{r}\mathrm{d}\phi.a_z=i_{\text{envolvida}} \tag{2.12}$$

$$\vec{a}_\phi.\vec{a}_z = |a_\phi|.|a_z|.\cos(90^\circ) = 0 \tag{2.13}$$

$$H_\phi = 0 \tag{2.14}$$

When removing one of the rheostats the equivalent resistance of the circuit in Figure 2.5, increases, as the electric voltage is constant it implies a decrease in the electric current, that is, there is a lower density of magnetic field lines around the wire, justifying a lower sensitization in the sensing coil and consequently the measurement of a lower induced voltage.

2.5 Final Considerations

The transformer is an electrical device whose characteristic can be isolating, lowering or raising the electrical voltage depending exclusively on the relationship between voltage and the number of turns of its windings.

The air gap is the essential component in the efficiency of such characteristics of transformer operation, being necessary for the best magnetic coupling between the windings.

The replacement of the conventional multimeter by the clamp meter, in alternating current circuits, with small deviations in measurement and with no need to open the circuit.

The impossibility of using the clamp meter in direct current circuits.

The dependence of the induced voltage in the sensing coil on the direction of the magnetic field passing through it. It can be seen that there is an induced voltage only when the magnetic flux is changing, thus confirming Faraday's Law.

There is a magnetic field for any circuit, fed by any source, according to Ampère's Law. The relationship between electric current and magnetic field is observed, and the reverse is true.

2.6 References

[Electromagnetism, pg.189-190, pg. 207, LTC, Rio de Janeiro, 1983.

[2] EDMINISTER, J.A., **Electromagnetism**, pq. 1, pg. 130-132, pg. 184, McGraw-Hill, São Paulo, 1980.

[3] HALLIDAY D., RESNICK R. WALKER J., **Fundamentals of Physics Electromagnetism** 4^a, pg. 183, pg. 190, pg. 207-212, pg.216-217, pg. 238, LTC, Rio de Janeiro, 1996.

[4] REITZ, J.R., **Fundamentos da teoria eletromagnética** 3^a, Campus, Rio de Janeiro 1988.

CHAPTER 3: HYSTERESIS CURVE

3.1 Summary

The use of ferromagnetic materials, in transformers or information storage, is specifically due to the magnetization properties described through the hysteresis curve. The use of these materials, where their magnetic permeability characteristics, ferromagnetic domain and all the phenomena related to their crystallography, when subjected to a magnetic field represent a hysteresis curve.

The energy variation in magnetic systems is analyzed through the Faraday-Lenz Law, using the interpretation of dissipated energy, which is visualized in the hysteresis curve. Experiments were performed using this knowledge and the operating principle and phenomena in the ferromagnetic core of the transformer caused by the effects of the hysteresis phenomenon were explained.

Using the transformer with two wire windings, which wrap around the arms of a metal frame and an electric current applied to one of the windings, with a resistor in each winding and a capacitor in series to the secondary winding. It is observed, the phenomenon of inductance, the voltage drop in the secondary winding and a smaller dispersion of the magnetic flux.

Therefore, the relations of the ferromagnetic characteristics of the material used in the core of the transformer attribute in a material easily magnetizable and demagnetizable, where the exchange of its magnetic domains favours for lower power dissipations.

3.2 Introduction

Magnetism is the phenomenon where materials impose an attractive or repulsive influence on other materials, which material is called magnetite.

Currently, many technological devices depend on magnetism and magnetic materials, such devices include electric power generators and transformers, electric motors, radios, televisions, computers and components for sound and video production systems. Some examples of materials that exhibit magnetic properties are transition metals such as iron (Fe), carbane monoxide (Co), nickel (Ni) and gallium (Ga).

However, all substances are influenced to a greater or lesser degree by the presence of a magnetic field, where through the hysteresis cycle, you know the relationship between magnetic force and the resulting induced magnetization in a substance, because the macroscopic magnetic properties of materials is a consequence of the magnetic moments that are associated with individual electrons, where each electron in an atom has magnetic moments that have their origin from two sources, one related to its orbital motion around the nucleus and the other along the axis of rotation.

As certain metallic materials possess a permanent magnetic moment in the absence of an external field and manifest very significant and permanent magnetizations, they are called ferromagnetic materials [1, 2, 3].

The relation in which a magnetic field $\vec{H}$ varying in time can produce an electric current in a closed circuit is related by the Faraday-Lenz Law [1, 3].

$$f.e.\mathrm{m} = -\int \frac{\partial B}{\partial t}.\mathrm{ds} \tag{3.1}$$

Therefore, it is called the transformer induction equation. As ferromagnetic materials acquire significant magnetization when subjected to a magnetic field, generated, for instance, in the coils of a transformer as shown in Figure 1.3

When considering the electric circuit in Figure 1.3, connecting the primary coil to a sinusoidal electric voltage, the effect of an applied field $\vec{H}$ on the magnetic induction $\vec{B}$ in the ferromagnetic metal, will be given through work done against the induced m.f.e. in the coil where [1,3]:

$$dW_b = H.dB \qquad (3.2)$$

The variation in the magnetic flux direction, due to the variant electric voltage, performs a process of demagnetisation and magnetisation, as it modifies the magnetic domains of the ferromagnetic core, and throughout this cycle the hysteresis phenomenon occurs in this material, where the hysteresis curve characterises the loss of energy in the form of heat when performing work under adverse phenomena to undertake the magnetisation and demagnetisation of this ferromagnetic core [1,2,3].

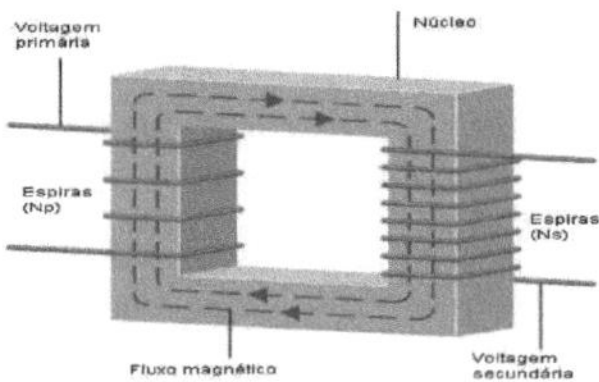

Figure 1.3 - Diagram of a transformer with an iron core and the magnetic flux lines. Source: Own authorship, 2005.

3.3 Experimental procedure

We measure in the circuit, as shown in Figure 2.3, with the oscilloscope, the electric voltage on the 47 Ω resistor, which is in Figure 3.3 and on the 330 nF capacitor represented in Figure 4.3.

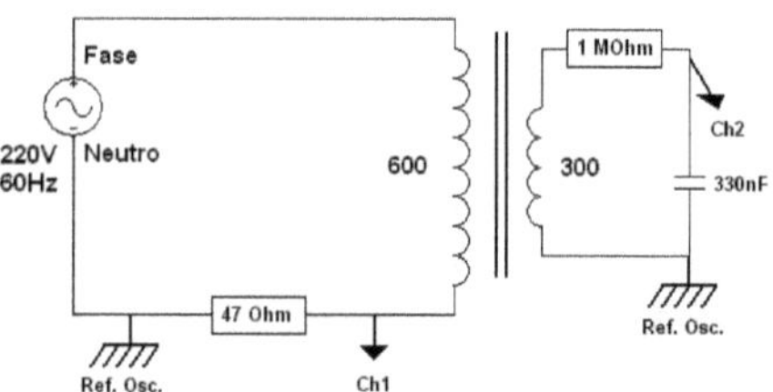

Figure 2.3 - Schematic of the circuit used to measure the hysteresis curve. Source: Own authorship, 2005.

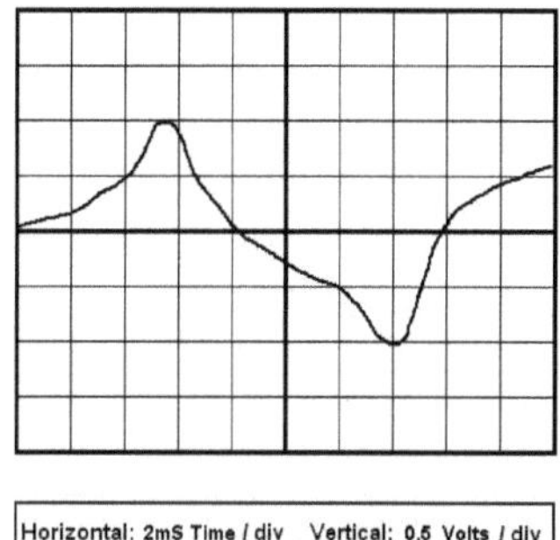

Figure 3.3 - Electric voltage on the 47 Ohms resistor. Source: Own authorship, 2005.

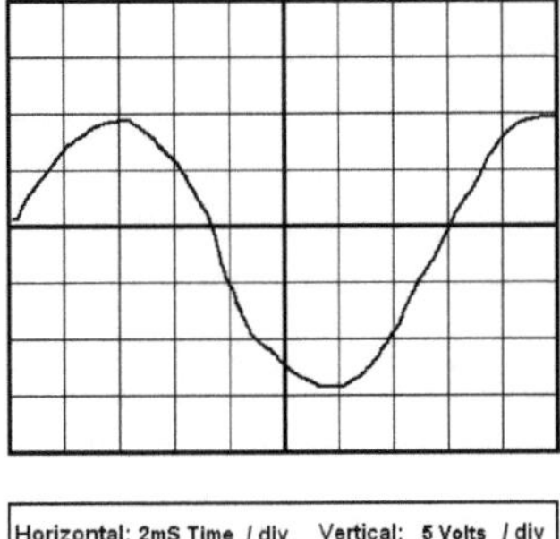

Figure 4.3 - Electrical voltage on the 330 nF capacitor. Source: Own authorship, 2005.

Channel 1 and 2 were placed in GND and the "XY" mode on the oscilloscope was activated, then the point was centred on the screen and the channels were removed from GND to activate AC mode, then what was observed was drawn on the oscilloscope screen noting the voltage value when x=0 and y=0, as shown in Figure 5.3.

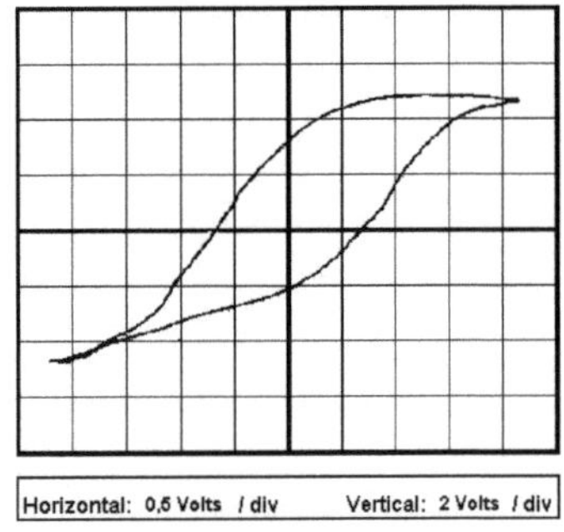

Figure 5.3 - Oscilloscope screen showing the visualization of channel 1 and 2 in XY mode. Source: Own authorship, 2005.

3.4 Results and discussion

The maximum electrical voltages on the capacitor, and the 47 Ω resistor, measured with the oscilloscope and presented in Table 1.3, represents two electrical quantities, V_R and V_C . In channel 1, presents the value V_R that designates the magnetization current [5], and originates from the orbital rotation movements of the electrons attached to the atomic nuclei, each generating its own magnetic field, which form magnetic dipoles that react with the presence of an external magnetic field $_{Bexterno}$, if this reaction is in the direction of weakening the external field, the material is diamagnetic, otherwise, the material is ferromagnetic and if there is no reaction, the material has no relevant magnetic effect [4].

Therefore, in channel 1, the voltage of the 47 Ω resistor, an electric voltage is observed with an effect proportional to the magnetic field $_{Haplicated\ to}$ the ferromagnetic core.

Table 1.3 - Maximum voltages in capacitor and resistor.

Maximum Voltage (Volts)		Error (%)
V_C	2,4	0,05
V_R	19	0,5

Source: Own authorship, 2005.

In channel 2, presents the V value$_c$ of the RC circuit connected to the secondary coil as presented in Figure 2.3, because of the high value of the resistance (1 MΩ), the time constant of the RC circuit is high enough for the sine to be fully seen, eliminating possible discrepancies, that is, the capacitor acts as a filter.

Thus, the signal of channel 1, as presented in Figure 3.3, shows a distinct form of the applied sinusoidal signal, because it describes the effect of magnetization on the core, that is, the effect of energy losses occurring in the ferromagnetic core and in channel 2, as shown in Figure 3.4, one can notice a difference between the signal of channel 1, because there is a filter, although in the transformer circuit, ideally, only the variation of the amplitude of the input signal relative to that of the output should occur.

Using the oscilloscope the electrical voltage V_y (x=0) and V_x (y=0) was measured, as shown in Table 2.3. For the hysteresis loop, Figure 5.3, to be observed in the oscilloscope, a transformer is required whose iron core reaches saturation at the extremes of the sinusoidal voltage applied to the circuit, therefore, in channel 1 (horizontal input) a voltage proportional to the H-field should be connected and in channel 2 (vertical input) an electrical voltage proportional to the B-field, being high value of the resistance of 1 MΩ, in the circuit of Figure 1.3 necessary to consider the circuit (RC) open, so the magnetic field H in the iron will only depend on the electric current in the primary [4].

Table 2.3 - Electrical voltages V_y (x=0) and V_x (y=0).

Voltage (Volts)		Error (%)
V_y	0,7	0,05
V_x	2,4	0,2

Source: Own authorship, 2005.

When y=0, the magnetic magnitude V_x represents the coercive force H_c , which is responsible for reducing the remaining flux density, which generates losses in the transformer and the lower this force H_c , the easier it is to magnetize and demagnetize the material. When x=0, the magnetic magnitude, V_y , represents the remaining flux density B_r [1,3,4,6].

Thus, when a material is saturated and reduces the H field, the hysteresis effects start to appear and it cannot come back through the same magnetization curve, even after H=0, B=B_r , and when H is reversed, then the density decreases to zero, thus the cycle is completed obtaining the hysteresis loop, Figure 6.3, which presents all the quantities involved in hysteresis cycle.

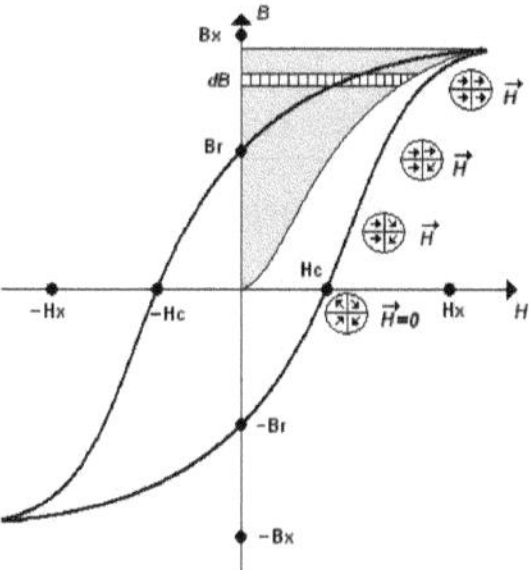

Figure 6.3 - Hysteresis loop, the coercive force Hc and the remaining flux Br. Source: Own authorship, 2005.

Iron when subjected to the varying magnetic field heats up for two reasons: eddy currents and the friction in the inversion of magnetic dipoles, which are the hysteresis losses. These losses are proportional to the area of the hysteresis loop. The energy that is delivered to the transformer can be written in terms of equation 3.3 [4]:

$$W= \int vidt = \int N \frac{d\varphi}{dt} idt = \int Nid\varphi =$$
$$\int N \frac{Hl}{N} d(BA) = Al \int HdB$$

(3.3)

So, this energy can be obtained by multiplying the shaded area in Figure 6.3 by the volume Al of the iron. Part of this energy stored in the magnetic field is transformed into heat, this heating is null if it reduces the electric current and the round-trip path in the BH curve, coincide, however, since it does not coincide, the hysteresis loop is proportional to the energy of the source, which feeds the circuit of Figure 2.3, which is transformed into heat, in each cycle, that is, in the magnetization and demagnetization of the ferromagnetic core. [4]

The area of the square in Figure 5.3 was calculated:

$$A=0.5 \times 2 = 1u.a$$

(3.4)

By inspection it was seen that the area under the graph of the hysteresis curve consists of approximately 8 squares (Figure 5.3). Since the area of each square is equal to one, and the energy is proportional to the amount of heat, then $Q \approx 8 (Joules)$.

The shape of the hysteresis curve for ferromagnetic materials is of considerable practical importance, as the area within a cycle represents the order of magnitude of work required for magnetisation or demagnetisation of the ferromagnetic material [1,2].

These materials are classified as easily magnetizable or hardly magnetizable, where the terms soft or hard are used respectively, based on their hysteresis curve. There are several effects that generate energy losses in ferrosilicon magnetic materials; the eddy currents, i.e., transient electrical voltage gradients that originate eddy currents; the magneto-anisitropy energy, i.e., the effect of the different crystalline orientation of the grains, and thus decreased magnetic permeability; the magnetostriction, that is, the variation of the bond distance between the atoms, when the electrons suffer rotation during the magnetization, where the fields of the dipoles originate repulsion and attraction, therefore, leading to contraction or expansion of the metal, therefore generating the noise of the transformers and loss of energy.

Soft magnetic materials, whose hysteresis curve is narrow, i.e. presenting a small coercive force, are used in devices subjected to magnetic fields with large alternation rates, where energy losses must be low, e.g. transformer cores [2].

Consequently, this material has a high initial permeability, in addition to a low coercivity to reach its saturation magnetization with the application of a relatively small field. Iron alloys with 3% to 4% silicon are the most widely used soft magnetic materials, but before 1900 low carbon steels were used in low frequency power devices such as motor transformers and generators with high core losses.

However, when exceeding 4% silicon, a decrease in iron ductility, saturation induction and Curie temperature occurs, i.e. temperature at which the ferromagnetic material becomes paramagnetic [2].

However, by using laminated structures, "stacked plates", a decrease of eddy current losses in transformer cores can be achieved, and stacked one on top of the other with a thin insulating layer in between. Another decrease of energy loss in transformer cores was achieved in the 40's with the production of grain oriented ferro-silicon plates, by combination of cold working and recrystallization it was possible to produce a 3% iron (Fe) plate with grain orientation (001), which is a direction of easy magnetization when subjected to a field applied in the direction parallel to the rolling of the plates, increasing the permeability and lower hysteresis losses than Fe-Si with random texture [2].

The hard ferromagnetic materials whose hysteresis curve is wide, i.e. presenting a large coercive force, are used in devices subjected to magnetic field with low rates of alternation of demagnetisation or magnetisation, i.e. the need to fire high energy for the process, enables the magnetic permanence of the substance, after the magnetic domains, to be oriented by the magnetic field, for example, in floppy disks ½, ¾ inch, audio and video magnetic tapes.

Meanwhile metallic glasses is a recent class whose dominant feature is a non-crystalline structure, consists unlike metals, of various combinations of the ferromagnetic Fe, Co and Ni as the metalloids and Si, being exceptionally soft, the applications of these metals refer to cores of transformers of small power loss, magnetic sensors and recording heads [2].

3.5 Final Considerations

The hysteresis loss is proportional to the frequency, in circuits subject to alternating current. The internal area of the hysteresis ring measures the energy dissipated, or work done, during the magnetisation and demagnetisation cycle.

A varying magnetic field produces transient electric voltage gradients that give rise to eddy currents or eddy currents inside the ferromagnetic material.

The total magnetic energy of a ferromagnetic material is the sum of the energetic contributions of several phenomena; of exchange of magnetic domains, magnetostatic, magnetostrictive, magnetocrystalline. There are several beneficial effects of using soft magnetic materials in the transformer core such as reduction, of eddy currents, magneto-anisitropy energy, and magnetostriction.

The term soft refers to a narrow hysteresis curve, i.e. low energy loss in the variation of the material domains, i.e. easily magnetisable and demagnetisable.

Metallic glasses constitute a recent class being exceptionally soft, among the applications we can refer to transformer cores of small power loss, magnetic sensors and recording heads.

One can analyse and observe the hysteresis curve through a circuit involved with a step-down transformer 220V/12V, resistors, a capacitor and voltage source, as well as analyse the magnetic properties of the hysteresis curve, analyse electrical and magnetic magnitudes through visualisation in an oscilloscope. It was also observed the working principle and concept of transformers, application of mathematical formulas and hysteresis curve.

3.6 References

[1] REITZ, J.R., MILFORD, F.J., CHRITSY, R.W, **Fundamentos da Teoria Eletromagnética,** 3ª ed.

[2]   SMITH, W.F., **Principles of materials science**, 3rd ed. pg. 671-681, Mc Graw-Hill, Amadora, 1998.

[3]   KRAUS, J.D., Carver, K.R., **Electromagnetism,** 2nd ed. pg. 187, pg. 211-220, pg. 222, Guanabara, Rio de Janeiro, 1978.

[4]   QUEVEDO C. P., **Electromagnetism**, pg. 353-358, pg. 424-425, Loyola, São Paulo, 1993.

[5]   JORDÃO R. G., **Transformadores,** 1ª ed., pg. 12-14, Edgard Blücher, São Paulo, 2002.

[6]   HAYT Jr, W. H., **Electromagnetism,** 3rd ed., pg.249-250, LTC, Rio de Janeiro, 1983.

CHAPTER 4: OSCILLATORS

4.1 Summary

This chapter discusses an oscillator circuit, which consists only of capacitor and inductor, which has as its utility the tuning of transmitters and receivers. It shows the importance of inductors in an oscillator circuit, both in transmission and reception, where the transmitter-receiver circuit is 100% in tune, with a certain quality factor (Q), that is, when the capacitance reactance (X_C) is equal to the inductive reactance (X_L) forming an angle of 0°, that is, when the natural frequency of the circuit is equal to the oscillation frequency.

When an alternating e.m.f. is applied, this phenomenon is called resonance, since any difference will cause interference to the circuit. It was also proven that the nominal value is acceptable with the value read on the multimeter of the capacitances and that the electric voltage decreases with the distance between the coils, so the focus of the experiment is the assembly of a tuned transmitter-receiver system, a tuned LC oscillator circuit and the tuning process of the transmitters and receivers.

4.2 Introduction

The ideal LC circuit, composed of inductor and capacitor, generate oscillations in the same way as in the mass-spring system, where the Law of Conservation of Energy is observed. In the combination of these two elements the energy stored in the electric field of the capacitor flows into the magnetic field of the inductor, and this energy stored in the magnetic field of the inductor, in turn flows into the capacitor, i.e., remaining indefinitely as an energy exchange between the electric field of the capacitor and the magnetic field of the inductor.

It happens that, the load, the electric current and the potential difference vary in a sinusoidal way, that is, the circuit oscillates with a frequency W. However, in the real LC circuit the oscillations are gradually damped, because there is a resistance that dissipates energy, and the oscillations are eventually extinguished.

However, if you sustain the electromagnetic oscillations for a certain frequency in an LC circuit, called natural angular frequency Wo, and approach another LC circuit whose frequency is also Wo, the phenomenon of mutual inductance will occur between the two circuits at resonance, but a maximum of energy transfer will occur only at this resonant frequency Wo.

Therefore resonance becomes a very important issue in engineering designs, for example in radio receivers in air traffic control systems, where the values of the resonant frequencies of the receiver circuit are adjusted to the natural frequency of the radio station designed to pick it up, i.e. the energy of the Wox (out-of-resonance) frequencies of the other stations is lower than that produced by the frequency of the station to which the circuit is tuned.

Analysing the oscillations of an ideal LC circuit the energy U present at any instant t in the circuit will be:

$$U = U_B + U_E$$
$$U = \frac{1}{2}Li^2 + \frac{q^2}{2C} \tag{4.1}$$

Since U is constant, its derivative is zero, thus:

$$\tag{4.2}$$

Where U_B and U_E are respectively the magnetic and electric energy, Q is the amplitude of the variations of the charge and W is the angular frequency of the magnetic oscillations. We have that:

$$W=\frac{1}{\sqrt{LC}}$$
$$U_E = \frac{Q^2}{2C}\cos^2(wt+\theta)$$
$$U_B = \frac{Q^2}{2C}\text{sen}^2(wt+\theta)$$

(4.3)

As presented in Figure 1.4, the maximum values of U_B e U_E are the same, and when one reaches a maximum the other is zero.

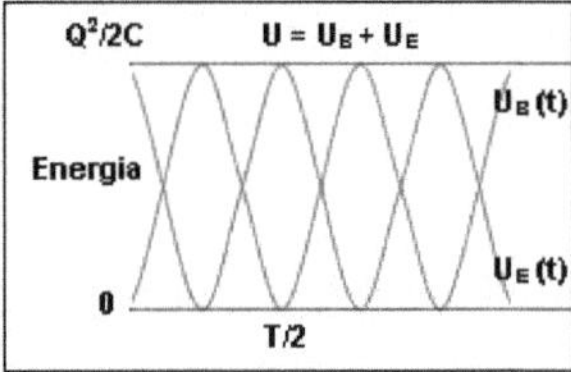

Figure 1.4- Magnetic and electric energy stored in the circuit. Source: Own authorship, 2005.

4.3 Experimental Procedure

The capacitance of the capacitors was measured by the multimeter and compared with the nominal value thus having a standard deviation. The circuit in Figure 2.4 is assembled.

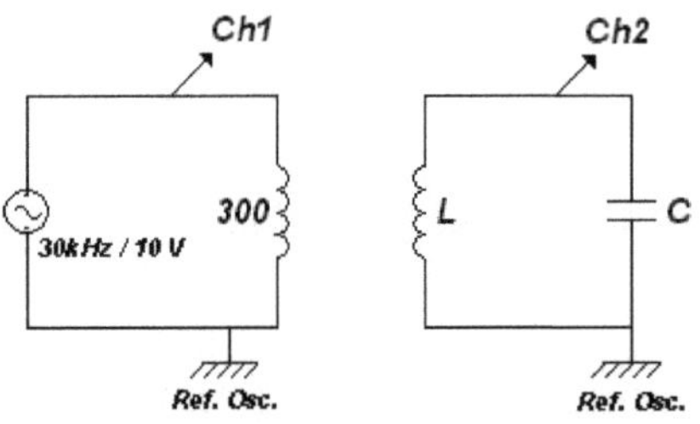

Figure 2.4 - Circuit for experiment. Source: Own authorship, 2005.

The oscilloscope traces were set, in GND mode, so that channel 1 and 2 were, respectively, 2 lines above and 2 lines below the central axis and CH $=_1$ 5 V/div, CH_2 = 1 V/div, initially.

The function generator was set to produce a sine wave with amplitude of 10 V_{pp} and frequency of 30 kHz. The coil of the generator was placed at a distance of 3 cm from the coil of the LC circuit. Adjustments were made, on the oscilloscope, so that the signal on channel 2 was readable, then the frequency, on the function generator, was varied and what was observed on the oscilloscope on channel 2 was reported.

Then the value of the frequency passing through a maximum point was calculated and their respective frequencies were noted and the inductances for each LC set were calculated according to Table 2.4. With these values the average value of L1 and L2 was found.

In the circuit of Figure 2.4, now with a distance of 1 cm between the generator coil and coil L1, the value of the phase between the voltage in the generator and the voltage in the LC circuit was measured with an oscilloscope. The frequency was varied around the resonance value and the behaviour of the phase as a function of the frequency was observed. With this, the frequencies corresponding to 70% of the value of the electric voltage on the LC circuit were found, and then the quality factor Q was calculated. The frequency was varied to its resonance value and the values in Table 3.4 were filled in.

4.3 Results and Discussions

The variation of the nominal values read was accepted according to the technical standards, which estimate for the capacitor with tolerances below 10%, which was verified as shown in Table 1.4

Table 1.4 - Nominal and read capacitor values.

Nominal Value (nF)	Read Value (nF)	Variance % Deviation
C1 = 10	10.6 ± 0.1	5.66
C2 = 22	22.4 ± 0.1	1.78
C3 = 33	34.2 ± 0.1	3.51
C4 = 330	3.2 ± 0.1	0.60

Source: Own authorship, 2005.

The variation in the generator frequency on channel 2 corresponds to a variation in the amplitude of the electric voltage, existing after passing through a point a dual relationship of proportionality, that is, the increase in frequency or its decrease in relation to the variation in voltage amplitude is relative to a midpoint, i.e. if the frequency is approaching this point, either because it is decreasing or increasing, there is an increase in the electric voltage, when it was closer to this point.

The frequency of this point (maximum point) is the resonance frequency of the circuit. The value of this frequency, measured on the function generator, is 41.2 kHz. For the LC circuit in series there is an increase in the resonant voltage, which is the ratio of the maximum stored energy at resonance to the resonant frequency. Therefore, a maximum energy transfer occurs at the frequency point where the voltage passes through a maximum value.

The value of the inductance, average and of each circuit, is calculated by the expression: $F_0 = \frac{1}{2\pi\sqrt{LC}}$, whose values are shown in Table 2.4:

Table 2.4 - Frequency values measured with each LC array

LC Set	Frequency (kHz)	L(mH)	L(mH) medium
C1-L1	41.2 ± 0.1	1.493	
C2-L1	28.6 ± 0.1	1.409	
C3-L1	23.4 ± 0.1	1.403	1,417
C4-L1	7.5 ± 0.1	1.365	
C1-L2	12.5 ± 0.1	16.22	
C2-L2	8.5 ± 0.1	15.95	
C3-L2	7.0 ± 0.1	15.68	14,095
C4-L2	3.0 ± 0.1	8.53	

Source: Own authorship, 2005.

When the distance between the generator coil and the LC circuit coil was changed to 1 cm, the value of the resonance frequency found had the same value when the distance was 3 cm.

The phase between the generator voltage and the voltage on the LC circuit is 187.8° as per the result of equation 4.4 and 4.5, which was calculated using a simple rule of three through the number of divisions of the oscilloscope time axis.

$$23 \text{ divisions - } 360°$$

$$(4.4)$$

$$12 \text{ divisions - } X° \qquad (4.5)$$

$$X° = 187,8°$$

$$(4.6)$$

The value of the voltage on the LC circuit, at resonance frequency is 14 ± 0.05 V. When varying the frequency, the behaviour is to go out of phase, so the phase goes from 0° to 187.8°, taking as reference the signal of channel 1 (Ch1) of Figure 2.4.

The energy stored in the electric field of the capacitor flows into the magnetic field of the inductor, and this energy stored in the magnetic field of the inductor, in turn flows into the capacitor, that is, maintaining indefinitely as an energy exchange between the electric field of the capacitor and the magnetic field of the inductor. If you strengthen the electromagnetic oscillations for a given frequency in an LC circuit and approach another LC circuit with the same frequency, the phenomenon of mutual inductance will occur between the two circuits, but a maximum energy transfer will occur only at the resonant frequency (phase equal to zero).

The Q factor or quality factor of an LC circuit, as of others, is a qualitative measure of a system's resonance system. These respond at frequencies close to natural than at other frequencies. This factor is calculated by the formula:

$$Q = \frac{F_0}{(F_2 - F_1)} \qquad (4.7)$$

Where F_0 is the resonance frequency and $(F - F_{21})$ is the frequency range of the passband. It can be seen that as the Q factor increases the passband $(F_2 - F_1)$ decreases, i.e. the oscillator circuit becomes even more selective, so it is a measure of sharpness of the resonance peak, so the higher the Q the sharper the peak. However, Q is also the ratio of the maximum energy stored at resonance and the energy dissipated per cycle. So there is a greater amount of energy flowing in the circuit.

In the circuit of Figure 2.4, when we move circuit 1 (function generator in series with coil 300) away from circuit 2 (coil L2 in series with capacitor C), the mutual inductance phenomenon has its magnetic field density affected in the relationship between these circuits, although the resonance frequencies are the same and the power flow supplied by the alternating source to circuit 1 has remained unchanged.

Therefore as you move the circuits further apart, the density of the field lines decreases more and more, and the energy received by circuit 2 decreases, because the induced current is smaller, which is viewable by the voltage decrease as presented in Table 3.4 and Figure 3.4. However if there were an increase in the voltage in circuit 1, increase in the charge rate, electric current, to compensate for the distance, the effect would be less noticeable.

Table 3.4 - Voltage values as a function of the separation distance between coils

Distance (cm)	Voltage (V)
1	14.0 ± 0.1
2	8.4 ± 0.1
3	6.0 ± 0.1
4	4.4 ± 0.1
5	3.0 ± 0.1
6	2.4 ± 0.1
7	1.9 ± 0.1
8	1.44 ± 0.1

Source: Own authorship, 2005.

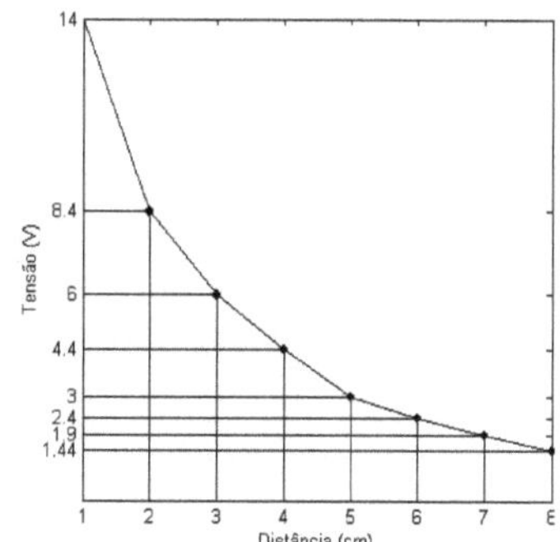

Figure 3.4 - Electrical voltage versus distance. Source: Own authorship, 2005.

4.4 Final Considerations

In this circuit, at high frequencies, the impedance of the circuit is dominated by the inductive term, consequently, at low frequencies, the impedance is dominated by the capacitive term.

The quality factor, Q, is a measure of the resonance peak, so the higher the Q, the sharper the peak.

The resonant frequency is the frequency at which the impedance of the LC circuit is real;

The maximum energy transfer, in the mutual inductance phenomenon between the two resonant circuits, will occur only at this resonant frequency W .0

4.5 References

[1] RESNICK, R., HALLIDAY, D., WALKER, J., **Fundamentals of physics, electromagnetism,** 4th ed. LTC, Rio de Janeiro, 2000.

[2] TIPLER, P. A., **Física, eletricidade, magnetismo e ótica,** 4^a ed., pp. 289-292, LTC, Rio de Janeiro, 2000.

CHAPTER 5: ELECTROMAGNETIC WAVE TRANSMITTER

5.1 Summary

This chapter covers the experiment on electromagnetic wave transmitter circuit showing the process of tuning an AM radio station, through the use of LC oscillators, are generated by oscillating electrical charges that propagate in various media, crossing any kind of obstacle, including vacuum.

Audio and video waves contain information, but because they are low frequency they cannot propagate in space by themselves, thus requiring carrier waves (high frequency waves). The experiment shows the behaviour of the waves on the oscilloscope in several different situations, observing that the carrier wave does exist, as well as the isolated audio wave when it leaves the transmitter's signal input.

It is concluded that by assembling the electromagnetic wave transmitter circuit it results that modulation is a process by which it modifies the radioelectric signal initially generated before it is radiated to transmit a message.

5.2 Introduction

In Italy, 1896, the English scientist James Clerk Maxwell when studying the electromagnetic nature of light, realised the theory on certain wavelike phenomena produced by the vibratory movement of electrons. From 1879, the German physicist Heinrich Hertz performed several practical experiments in which he uses an oscillating circuit or oscillator, a device in which the electric current circulates now in one direction and vice versa, whose phenomenon happens when there is an electric current running in a conductor producing an electromagnetic field, known as autoinduction.

Hertz's experiments were the starting point for the discovery of radiotelegraphy, radiotelephony and television and opened horizons for other physicists such as Eduard Branly and Guglielmo Marconi, resulting in a patent, 1896, for the invention of the wireless transmitting apparatus.

Humans can hear only in the frequency range of 20 Hz to 20 kHz, so the sound waves of the human voice must be transmitted in electromagnetic waves and then converted to be audible. The first role is played first by the microphone and the second by the loudspeaker, with the aid of valves (obsolete devices) or transistors (modern devices).

To avoid interference transmitting stations frequencies are regulated by governments, so the hertzian waves, also called electromagnetic waves, correspond to a range of frequencies and amplitudes. In general, commercial broadcasters use the range above 1000 meters wavelength, they are called long waves. In turn, the wavelengths between 100 and 1000 metres correspond to medium waves.

Short waves, from 10 to 100 metres long, are used for various purposes. Among the applications are police communications, aviator guidance in flight, radio amateurism and intercontinental programmes. Long waves differ from short waves in that they can go around large objects such as buildings, saws and others, while short waves require less energy and can be centred in thin beams.

These beams are used in the guidance/detection of various transports such as aircraft and submarines, radar waves are included. The waves used for high frequency radio and television come from the English term Very *High Frequency* (VHF) with wavelengths less than 3 metres.

When an electric current is forced to oscillate in another transmitting antenna, electromagnetic waves are produced which propagate, externally, at a speed of 300,000 km/s. Each oscillation of the current produces the emission of a wave, whose frequency is expressed by the number of complete cycles per second.

The wavelength, expressed in metres, is consequently equal to the speed in metres per second divided by the frequency (cycles/s.) Radio waves vary in length from less than a centimetre to 20 km, the path of which these radiated waves lose energy to the earth's surface.

In modern transmitters the electrical oscillations are generated by means of thermo-ionic valves, which integrate inductance (usually a coil of wire) and capacitance (usually a capacitor of metal foils separated by air) circuits.

The electrical oscillations are merely carrier waves which, modulated, carry the information. The antenna radiates the high frequency waves with great diffusion through space. Short waves are suitable for directional transmission.

5.3 Experimental procedure

The circuit of Figure 1.5 was assembled, without the "signal" block, on the *protoboard* with attention to its respective electrical connections, mainly with the transistor terminals. The oscilloscope with Ch1 x 10 was connected to the transmitter input, with initially on the scales (1uS /div - 0.2 Volts /div) in order to obtain a good visualization.

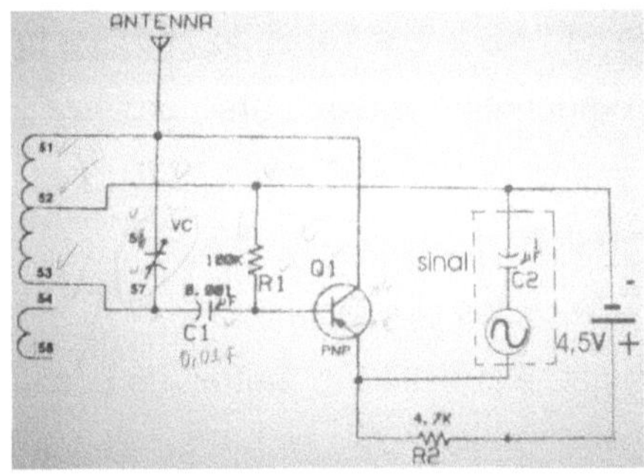

Figure 1.5 - Electrical schematic of the transmitter. Source: Laboratory Manual

The wave shape was observed and drawn, seen by Figure 2.5 and the frequency of the generated wave was measured. Channel 2 was connected to the generator, adjusting the signal to 300 mVpp with 500 Hz and the waveform was drawn, seen by Figure 3.5. The "signal" block was connected to the circuit as shown in Figure 1.5.

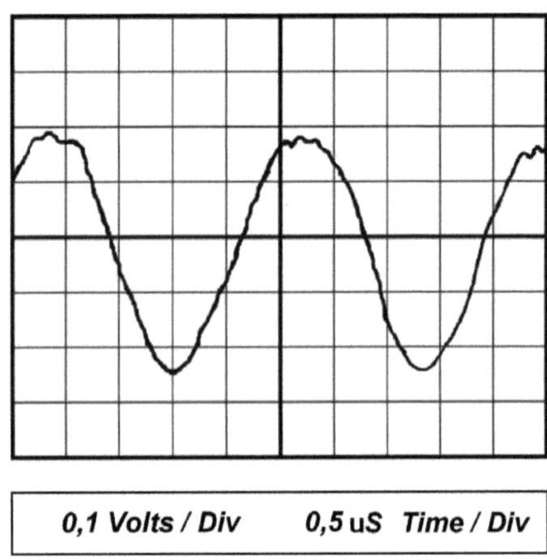

Figure 2.5 - Transmitted waveform when connecting Ch1 of the oscilloscope to the transmitter antenna.

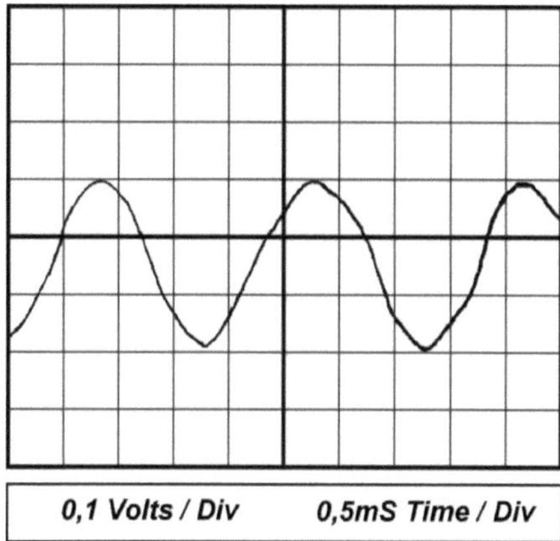

Figure 3.5 - Waveform shown on the oscilloscope when connected to the signal input on the transmitter.

A radio receiver was turned on and the tuning of the receiver was varied to a range, where no station is being received in order to adjust the transmitter so that signal reception occurs. With the oscilloscope, with Ch1 x 1 to the antenna, based on the time, around 0.5 ms/div that was, observed and drawn the transmitted waveform, seen in Figure 4.5. And consequently the frequency and amplitude of the generator signal was varied, which was observed and seen in Figure 5.5.

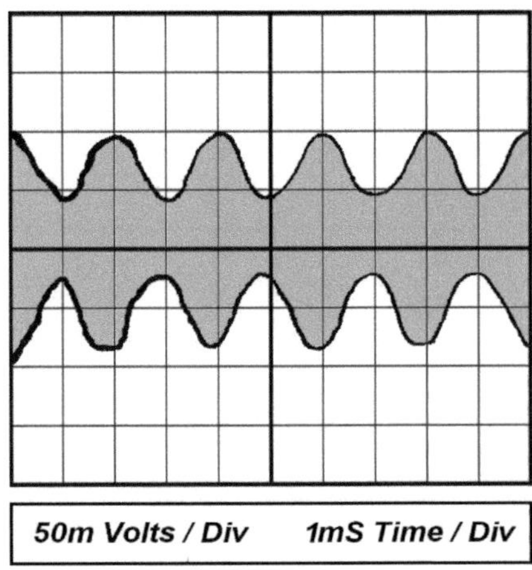

Figure 4.5 - Waveform shown on the oscilloscope when we adjust the transmitter to occur signal reception.

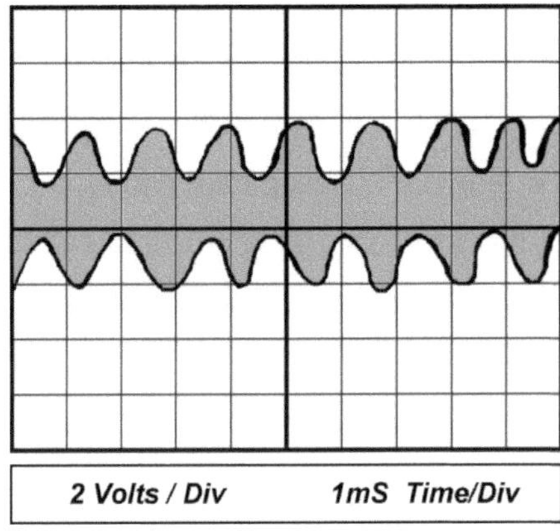

Figure 5.5 - Waveform when we vary the frequency and amplitude of the generator signal.

5.4 Results and Discussions

The waveform shown by Figure 2.5 is called the carrier wave. The waveform shown by Figure 3.5 is called the modulated wave. The combined audio and radio frequency oscillations are directed towards the transmitting antenna, from which they radiate in the form of electromagnetic waves. The radio frequency wave is called the carrier wave, when it carries audio signals, it is said to be modulated.

So, after we turn on the "signal" block the carrier wave started carrying audio signals, which turned it into a modulated wave. After that, we vary the tuning of the receiver to a range where it is not receiving any station and adjust the transmitter so that the signal is received.

Connect the oscilloscope to the transmitter antenna with the time base around 0.1 ms/div to observe the waveform in Figure 4.5. Now vary the frequency and amplitude of the generator signal and observe the waveform in Figure 5.5.

The oscillation has changes in the direction of its flow based on regular intervals, because oscillation is an electric current. In this sense, the transformation of sound waves into electromagnetic waves takes place, which when passed through the amplifier will be amplified.

In practice, the radio transmitter generates significant frequency oscillations, higher than those produced by the original sound source and which will radiate through space. As the amplifications of the high frequency oscillations which occur in the microphones pass to the modulator which is connected to the output of the oscillator, any increase in the electric current coming from the studio, as a result of sound variations, causes corresponding changes in the amplitudes of the wave contour of the oscillator resulting in amplitude modulation.

The electromagnetic waves radiated over the long distances result from the combinations of audio and radio frequency oscillations are directed to the transmitting antenna. Therefore, the carrier wave is the radio frequency wave when carrying audio signals is called modulated.

The transistor is an active element of the circuit generating the high frequency signals that are transmitted by the antenna. The frequency between 88 and 108 MHz is determined by L1 and CV. You can adjust the frequency at CV as explained above.

To maintain the oscillations part of the signal that appears at the collector is carried back to the emitter by capacitor C3. This is the feedback capacitor. The power of the transmitter and therefore its range which depends on the transistor. The electrical current that the transistor supports is fixed by R4 which must be 47 ohms or higher electrical resistance.

Higher value resistors may cause the transmitter to have less power, but smaller resistors cause currents that the transistor used cannot handle. Resistors R2 and R3 polarise the base of the transistor, i.e. they set the voltage that the transistor needs to work.

The function of capacitor C2 is to "decouple" the base of the transistor, i.e. to provide a path for the high frequency generated by the circuit so that it does not affect the bias circuit.

The electret microphone MIC has a transistor inside that needs to be polarised. This polarisation is done by R1 whose value is calculated as a function of the supply voltage. Converting the sounds into audio signals the microphone sends them to the base of the transistor via C1.

This so-called "coupling" capacitor, while allowing audio signals (sounds) to pass to the transistor base, prevents microphone bias from influencing transistor bias. The capacitors "isolate" the direct current circuits from polarization. The audio signals arriving at the base of the transistor influence the frequency of the generated signal, i.e. modulate this signal in frequency, thus producing a frequency modulated (FM) signal. By placing an antenna these signals can be transmitted into space.

To increase signal strength, one must separate the function of generating the signals from that of amplifying them. Thus, a small power transistor is used to generate the signals and then amplify these signals with transistors or stages until the desired final power is obtained for application in an antenna.

In amplitude modulation (AM), there is a circuit that produces variations in the intensity of the signal generated by the transmitter. The most common is that this variation is produced in the last or penultimate amplification stage. When a lower frequency signal amplitude modulates a high frequency carrier, the end result is not only a variation in the intensity of the transmitted signal. A "combination" of the signals also occurs with the production of new signals at the frequencies summing the difference.

This means that varying the frequency transmitted with the modulating signal, for example, in the case of voice transmission, varies in a range that corresponds to the modulating signal range. So if you modulate the AM signal over a 5 kHz band, it will vary by 5 kHz, up and down, then occupying a 10 kHz "width" band. This is why AM signals need a band of a certain width in the spectrum, i.e. a 10 kHz channel. This is the minimum separation that stations must have in order not to interfere with each other.

5.4 Final Considerations

From this experiment, the circuit transmitting electromagnetic waves, which are generated by means of electric charges propagating in space at the speed of light (300,000 km/s), was analysed and understood.

The atmosphere contains many electromagnetic waves of various frequencies which all reach receiving antennas, so every receiving device needs specific frequency bands to tune into the particular transmitted signal.

The waves that transport audio and video being of low frequencies would not be able to propagate without a carrier wave, because the audio and frequency waves take a "ride" to be transported to long distances, this process is called modulation.

When these waves reach a receiver it is called demodulated, i.e. the carrier wave is removed, as is the inverted audio signal, the receiver recovers only the non-inverted audio signal, and therefore identical to the signal emitted by the transmitter circuit.

5.5 References

[1] KOSOW, Irving L., **Electrical Machines and Transformers**, 6th edition, Editora Globo S.A., Rio de Janeiro - RJ, Brazil, 1986.

[2] NEWTON, C. Braga, **Curso Básico de Eletrônica,** pp 245-249, ed. Saber, São Paulo, 2001.

[3] Manual of the Electric Machines Laboratory. UNIFOR - CCT - 2004.

CHAPTER 6: RECEIVER OF ELECTROMAGNETIC WAVES

6.1 Summary

This chapter discusses the use of an oscillator circuit with variable frequencies to capture electromagnetic waves according to the properties described by James Clerk Maxwell's Theory.

The use of this theory quantifies the electromagnetic spectrum as infinite, so there are several oscillations with varying frequencies and amplitudes. Thus, using the electronic components: antenna, diodes, capacitors, resistors, inductors, an integrated circuit (IC) 324 and a headset, they make a circuit for the specific capture of a modulated electromagnetic wave, demodulation and conversion of electrical signal into sound.

The phenomena regarding interference and quality factor Q solidify for the variation of energy captured in the electromagnetic wave receiver system, because in the energy capture by the resonant circuit, there is a single frequency value for maximum power transfer and a range that corresponds to 70% of the captured signal.

Therefore, the phenomenon of the electromagnetic wave is observed as a means of transporting information and it is clear that the acquisition of a good receiver circuit requires sensitivity to small powers, high quality factor Q, amplification of the signal by converting it into sound to make it audible, when it comes to radio receivers where the information is sound.

6.2 Introduction

Dragging, i.e. the mutual phase of two pendulums was a phenomenon observed by Christian Huygens 300 years ago, in which two pendulums maintained a combined rhythm [1].

Thus, it took many discoveries in the field of electricity and magnetism to culminate in James Clerk Maxwell's theory of 1873 in which he succinctly stated the possibility of producing electromagnetic waves and the experimental proof by Heinrich Hertz in 1887 and the application of electromagnetic waves for sending telegraph signals by Guglielmo Marconi in 1895 [2,3].

The use of electromagnetic waves for the transmission of sound information occurred at the beginning of the 20th century, when Lee de Forrest and Reginand Ambrey Fessenden irradiated in the United States, singing numbers and violin solos, thanks to the invention of the triode, created by Lee De Forrest in 1906, which allows the rectification and amplification of electrical signals, enabling the hearing of complex sounds, transmitted by electromagnetic waves [4].

The transmission of an electrical signal of low frequency causes a large wavelength which needs another signal of high frequency to carry the information signal, by changing its characteristics, equivalent to the information signal.

The information signal is called the modulating signal and the high frequency signal is called the carrier wave, the result of the interference of one signal on the other is a third electrical signal called the modulated signal and the process involving the generation of this signal from the first two is known as modulation.

For the carrier wave there are several ways to change its amplitude and frequency characteristics, such as AM amplitude modulation, PM phase modulation, FM frequency modulation, and among others such as TDM, QAM, etc [5].

The atmosphere meets various electromagnetic waves of various frequencies and they all reach the radio receiving antennas where the electric field of the wave causes the free charges of the receiving antenna to move, which are forced to oscillate at the same frequencies as the electromagnetic wave oscillations.

The wave emitted by the radio station antenna reaches the receiving antenna where the signals from the broadcasters are picked up and then selected through a resonant circuit for each channel, and converted into sounds that we hear on our radio sets, Figure 1.6 shows a receiving circuit [6].

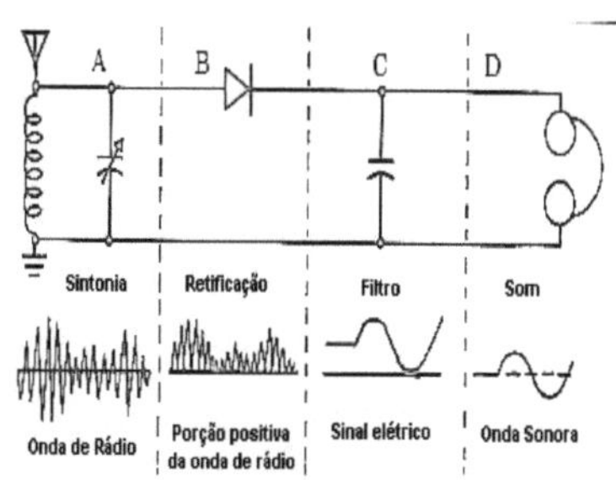

Figure 1.6 - Diagram of a galena radio receiver.

6.3 Experimental results

For this experiment, an electromagnetic wave receiver was assembled in an experiment case. The circuit was assembled as shown in Figure 2.6, then the earphone was placed in the ear and the tuning knob, variable capacitor, was varied until a station could be selected. An oscilloscope was used to observe the wave captured by the antenna, modulated signal, and then the carrier and modulating signal, as well as the amplification performed by the circuit.

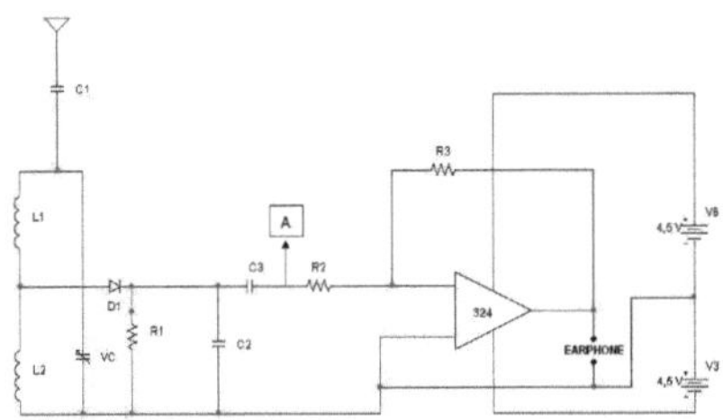

Figure 2.6 - Electrical scheme of the receiver circuit. Source: Own authorship, 2005.

6.3 Results and Discussions

Figure 2.6 shows a circuit consisting of a coil and antenna, with a variable capacitor associated on the secondary side, constituting a resonant circuit for station selection. The modulated radio frequency (RF) signals of the tuned station are applied between the plate and the cathode through the capacitor and the earphone.

The signal at point 52 of the circuit in Figure 2.6 shows the signal picked up by the antenna that has the carrier and the audio signal. While the signal shown by point A shows only the demodulated audio signal, thus it is possible to hear the sound transmitted by the station.

The diode presents the property of allowing the passage of electric current only in one direction, so use it to rectify the RF modulated signals in order to recover their envelope. Assuming that the diodes are not ideal, if the germanium diode is exchanged for a silicon diode, then the germanium diode will provide a voltage drop of 0.3 V, while the silicon diode will provide a voltage drop of 0.7 V.

Figure 3.6 represents the oscillation of a free charge on a transmitting antenna, this oscillation generates an electromagnetic wave that propagates in space, the oscillating electric field of the wave lies parallel to the antenna, in turn it propagates in a direction perpendicular to the antenna.

Figure 3.6 - Electrical wave generated by the oscillation of electrical charges in an antenna.

The polarization of a radio wave is determined by the position and direction of the electric field relative to the earth's surface as presented in Figure 4.6. The wave is horizontally polarised if the electric field lines of the radio wave are parallel to the ground and if the wave is vertically polarised it means that the lines are perpendicular to the ground.

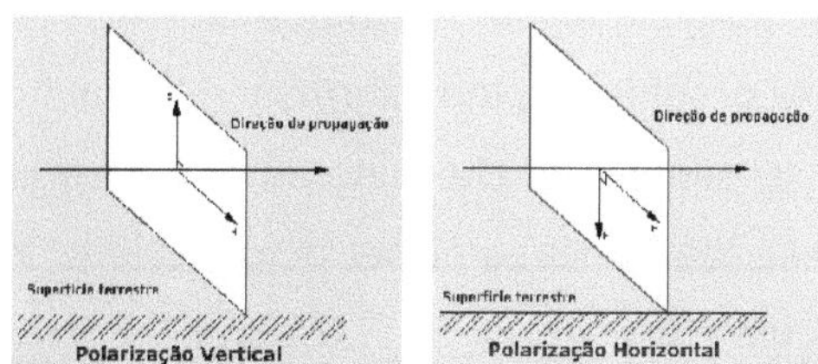

Figure 4.6 - Vertical and horizontal electromagnetic wave polarisation.

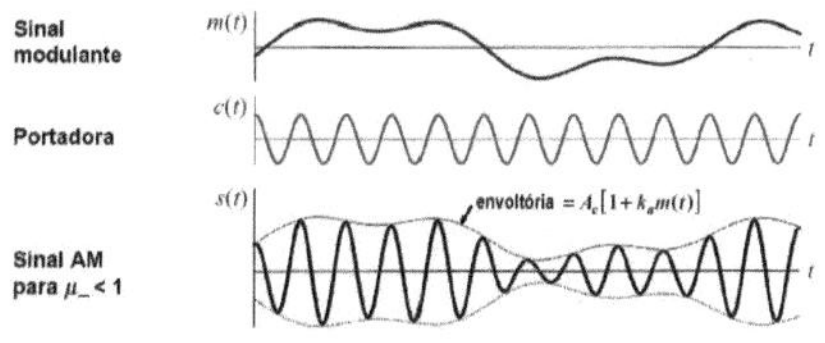

Figure 5.6 - Waveform of modulating, carrier, and modulated signals.

When the electric wave emitted by the radio station antenna reaches the antenna of a radio receiver the effect of the electric field of

61

the wave moves the free charges of the receiver antenna, which are forced to oscillate, at the same frequency as the oscillations of the wave.

This electrical current reaching the receiving antenna is small compared to the transmitting antenna, so the receiving circuitry amplifies it. Radio waves are generally referred to as radio carriers because they simply do the function of carrying energy to a remote server, where the data being transmitted is overlaid on the radio carrier, as presented in Figure 5.6 [8].

It turns out that for shortwave reception, radio transmission, polarization is not particularly important, because radio waves reflected by the sphere tend to lose their initial polarization. The recovered signal is composed of a portion proportional to the modulating signal and a DC level proportional to the amplitude received from the carrier.

As several transmitting stations send AM signals into space, the detector, does not select one among several stations and thus retrieves a practically unintelligible signal, so the basic AM receiver consists of a detector, capable of retrieving the information signal from the modulated signal, a selector, capable of choosing among several stations, and an amplifier, to make the retrieved signal perceptible to the human ear, as presented in Figure 6.6.

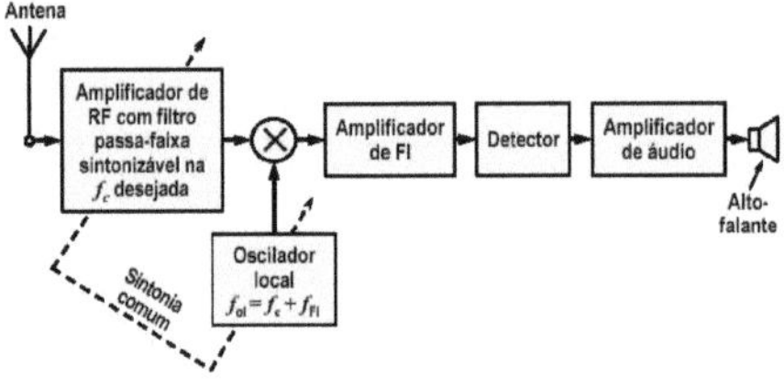

Figure 6.6 - Receiver block diagram.

It was observed that the two RFS amplification stages are tuned to a defined frequency, corresponding to the station you want to receive, making it necessary the band-pass filter composed of an LC circuit, in each of the RF amplifiers. The RF filter and the quality factor Q remain constant in the reception range, because in the RF range the frequency is high enough that the pellicular effect begins to appear, which consists of the passage of electric current through the periphery of the conductor, leaving its central portion without any function, this decreases the useful cross section of the conductor and its resistance will increase with increasing frequency [7].

We have that the quality factor Q is given by equation 6.1

$$Q = 2 \cdot \pi \frac{W_S}{W_D} \tag{6.1}$$

Where W_S is the maximum stored energy at resonance and W_D is the dissipated energy per cycle [5].

Therefore the receiver steps are to select which station to listen to through a tuning process, separate only one half of the radio wave which is used to extract the sound wave, separate the electrical signal equivalent to the sound wave you will hear and the process is completed through a component capable of transforming electrical signals into sound signals, the crystal headset.

6.4 Final Considerations

In the receiver, the integrated circuit (IC) amplifies the signal, the resistors form a resistive adder, the diode plays the role of a frequency synchronous switch and the set of inductors and capacitors form a tuned circuit that draws the full spectrum available in the switched signal.

The factors that affect the stability of the receiver are frequency and variations in temperature and humidity, as well as the transistor operating point.

The efficiency of an antenna depends on the correct relationship between its physical length and the wavelength of the signal it receives.

Advances in electronics popularised the use of radiotelegraphy and radiotelephony.

Each radio receiver has the means to select a frequency band, tune to a specific signal and converts the received signal into sound and amplifies it.

Demodulation is a process that consists in changing a characteristic of the carrier wave, proportionally to the modulating signal.

6.4 References

[1] KOSOW, Irving Lionel, **Electrical Machines and Transformers**, 8th Edition, Ed. Globo, 1972, p. 165-174, p. 230-242.

[2] **Manual de Laboratório de Máquinas Elétricas**, Fortaleza: UNIFOR, p. 47-57, 2005.

[3] IRWIM, J. D., **Circuit analysis in engineering,** 4th ed., Makron Books, São Paulo, p.588, 2000.

[4] GOMES, Alcides Tadeu, **Telecomunicações, Transmissão e Recepção AM-FM: sistema pulsados,** 14th edition, Editora Érica, São Paulo, 1998.

Buy your books fast and straightforward online - at one of world's fastest growing online book stores! Environmentally sound due to Print-on-Demand technologies.

Buy your books online at
www.morebooks.shop

Kaufen Sie Ihre Bücher schnell und unkompliziert online – auf einer der am schnellsten wachsenden Buchhandelsplattformen weltweit! Dank Print-On-Demand umwelt- und ressourcenschonend produziert.

Bücher schneller online kaufen
www.morebooks.shop

Printed by Books on Demand GmbH, Norderstedt / Germany